TC 3-22.9 Supplement

RIFLE GOLD BOOK

Table of Contents

This Supplemental is intended to expound upon the information found in TC 3-22.9.
The information has been provided by United States Army Marksmanship Unit and
has been approved for release by United States Army Infantry School within the
Maneuver Center of Excellence.
USAMU is part of the U.S. Army Accessions Brigade
and Army Marketing and Research Group.

CHAPTER 1 - FIREARMS SAFETY AND WEAPONS HANDLING

RULES OF FIREARMS SAFETY

1-1.   To effectively and safely handle weapons, Soldiers must apply the rules of firearms safety. These rules integrate the three components of weapons handling. They are designed to provide redundant safety measures when handling any weapon.

1-2.   This redundancy allows for multiple fail-safe measures to provide the maximum level of safety in both training and operational environments. A Soldier would have to violate two of the rules of firearms safety to injure or kill another.

Note. Unit SOPs, range SOPs, or the operational environment may dictate additional safety protocols; however, the rules of firearms safety are always applied. If a unit requires Soldiers to violate these safety rules for any reason, such as for the use of blanks, the unit commander must take appropriate risk mitigation actions.

1-3.   Any weapon handled by a Soldier must be treated as if it is loaded and prepared to fire. Whether or not a weapon is loaded should not affect how a Soldier handles the weapon in any instance. Soldiers must take the appropriate actions to ensure weapon status.

1-4.   Never point the weapon at anything you do not intend to shoot. Soldiers must be cognitively aware of the orientation of their weapon's muzzle and what is in the path of the projectile if the weapon fires. There are instances where this is unavoidable. When this occurs, the Soldier must minimize the amount of time his muzzle is oriented toward people or objects he does not intend to shoot, while simultaneously applying the other three rules of fire arms safety.

1-5.   Keep finger straight and out of the trigger well until ready to fire. Soldiers must not place their finger on the trigger unless they intend to fire the weapon. Mechanical safety devices on a weapon can fail and must not be solely relied upon for safe operation. Additionally, some weapons the Soldier may operate do not have any traditional mechanical safety. The Soldier is the most important safety feature on any weapon. The weapon is moved to safe when a target is not presented. If the weapon does not have a traditional mechanical safe, the trigger finger acts as the primary safety.

1-6.   Ensure positive identification of the target and its surroundings. The disciplined Soldier knows the target and what is beyond, in front of, and surrounding it. The Soldier is responsible for all bullets fired from their weapon, including the projectile's final destination.
Application of this rule minimizes the possibility of fratricide, collateral damage, or damage to infrastructure or equipment.

ADDITIONAL WEAPONS SAFETY CONSIDERATIONS

1-7.   The weapon is safe to operate. Just like other tools, rifles need regular maintenance to remain operational. Regular cleaning and proper storage are a part of the rifle's general upkeep. Never knowingly attempt to fire a damaged rifle. This could lead to serious injury. If there is any question concerning a rifle's ability to function, a knowledgeable armorer should look at it.

1-8.   You know how to use the rifle safely. Before handling a rifle, learn how it operates. Know how to safely open and close the action and remove any ammunition from the rifle or magazine. Get familiar with the basic parts. Using the technical manual, learn to disassemble and re-assemble the rifle. Remember that a rifle's mechanical safety device is never foolproof. Nothing can ever replace safe weapon handling.

1-9.   Use only the correct ammunition for your rifle.

1-10.   Wear eye and ear protection as appropriate. Rifles are loud and the noise can cause hearing damage. They can also emit debris and hot gas that could cause eye injury. For these reasons, shooters and spectators should wear safety glasses and hearing protection.

1-11.   Never use alcohol or drugs before or while shooting. Alcohol and other drugs are likely to impair normal mental and physical bodily functions. The combination of alcohol or drugs with guns is a dangerous mix.

1-12.   Practice safe weapons handling to make it habitual. Never take short cuts when it comes to safety. Insist those around you follow these rules. Be aware that certain types of weapons and many shooting activities require additional safety precautions.

1. Butt stock
2. Elevation wheel
3. Windage wheel
4. Ejection port
5. Front sight assembly
6. Barrel
7. Flash suppressor
8. Front sling swivel
9. Magazine
10. Magazine release
11. Pistol grip
12. Rear sling swivel
13. Butt plate
14. Hand guards
15. Carrying handle
16. Rear sight housing
17. Charging handle
18. Safety selector
19. Trigger
20. Bolt release and catch

Figure 1.1. Key components of the M16/M4 series of weapons

CHAPTER 2 - PRINCIPLES OF OPERATION

2-1.   The purpose of this chapter is to establish the procedures that follow to ensure that any firearm a Soldier picks up is unloaded before handling it further, and that the Soldier understands how to load the weapon. The following paragraphs describe the sequence in which to safely clear and load an M16 or M4 immediately upon acquiring positive control of the weapon, regardless of circumstances. The steps are listed below and explained in greater detail during the rest of the chapter.

2-2.   CLEARING PROCEDURES

a. Attempt to place the rifle on safe. The first step in clearing the M16/M4 is to attempt to put the rifle on safe. Remember that an M16/M4 that is cocked will allow the selector lever to rotate to the safe position, an uncocked M16/M4 will not. Simply stated, if the hammer is forward, the safety cannot be engaged.

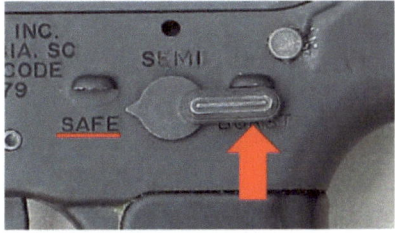

Figure 2.1. Place the weapon on safe

Figure 2.2. Remove the magazine

b. Remove the magazine. The second step is to remove the magazine by pressing the magazine release button. The Soldier presses the magazine release button and removes the magazine with the opposite hand.

Figure 2.3. Lock bolt to the rear

c. Pull charging handle to the rear, lock the bolt in place, and inspect the chamber. The third step ensures the chamber of the weapon is clear. To accomplish this, the Soldier points the weapon in a safe direction with the magazine removed and pulls the charging handle to the rear. Next, the Soldier locks the bolt to the rear by pressing the lower half of the bolt catch and slowly allow the bolt to go forward until it catches. Return the charging handle fully forward into the locked position. Finally, the Soldier visually inspects the chamber to ensure there is no brass or ammo remaining in the chamber area.

Figure 2.4. Inspect the chamber

5

d. Place the rifle on safe. The final step is to ensure the weapon is made safe after ensuring the clearance of the chamber. This is done by simply rotating the selector level to the "SAFE," position. This is possible because the rifle has been charged the selector lever can now be moved to the safe position. If you cannot place the rifle on safe, immediately notify the range safety.

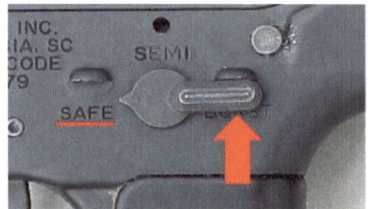
Figure 2.5. Place the weapon on safe

2-3.　LOADING THE WEAPON

a. While keeping the muzzle pointed in a safe direction ensure rifle is on safe, Figure 2.5.

b. Lock the bolt to the rear, paragraph 2-2-c, Figure 2.3. Check to make sure chamber is clear, Figure 2.4.

c. Insert a loaded magazine into the magazine well and firmly push upward until the magazine locks into position with an audible click.

d. Pull downward on the magazine to ensure it is locked in position.

Figure 2.6. Load magazine into rifle

Figure 2.7. Push in until click, then pull to check security

e. Depress the top portion of the bolt catch. The bolt will close and load the first round from the magazine into the chamber.

f. Close the ejection port cover.

Figure 2.8. Release bolt by pressing top of bolt catch

Figure 2.9. Close ejection port cover

6

2-4.   There may be circumstances due to designated force protection levels or direct fire control measures that require that a magazine is already inserted with the bolt closed on an empty chamber. Once the Soldier is either directed to charge his weapon or the presence of a threat emerges, he takes the following steps to safely and efficiently chamber the first round of ammunition from the magazine.

   a.   While keeping the muzzle pointed in a safe direction, ensure rifle is on safe.

   b.   Ensure the magazine is fully seated by pulling downward on the magazine. If it comes out, reseat magazine.

   c.   Pull the charging handle all the way to the rear and release. Allow the charging handle to go forward on its own.

Figure 2.10. Pull charging handle to the rear   Figure 2.11. Release charging handle

   d.   Close the ejection port cover.

2-5.   **FIELD STRIP AND INSPECTION OF THE M4/M16 SERIES RIFLE**

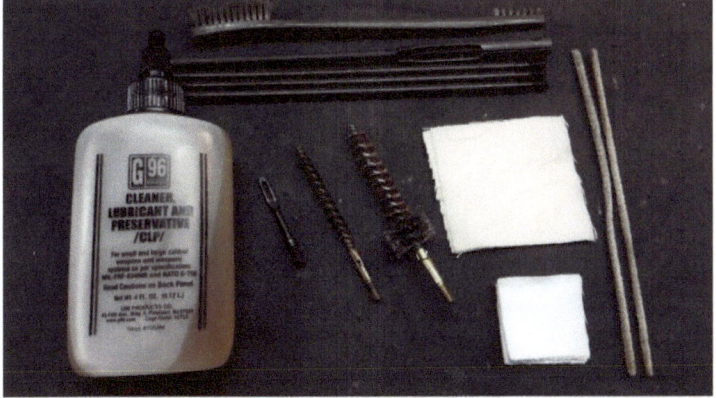

Figure 2.12. M16/M4 cleaning supplies

a. Inspect the weapon to ensure that it is clear by locking the bolt to the rear, move the selector lever to "safe" and visually inspect the chamber and bolt face to ensure there is no live ammunition in the weapon.

Figure 2.13. Clear the weapon and place on safe

Figure 2.14. Push out take-down pins

b. With the charging handle fully seated and in the forward position, press the bolt release and field strip the weapon by pushing out the front and rear take down pins.

c. Remove the bolt carrier group and charging handle.

Figure 2.15. Remove bolt carrier group and charging handle

Detent pin

Figure 2.16. Remove buffer and buffer spring

d. Compress the buffer tube detent pin controlling the buffer and buffer tube spring as it comes out of the buffer tube. Inspect the buffer spring for flat or worn spots and the buffer for excessive wear.

e. Bolt carrier group. Remove the firing pin retaining pin, the firing pin, the bolt cam pin and bolt.

Figure 2.17. Remove firing pin retaining pin

Figure 2.18. Disassemble bolt carrier group

f. Bolt carrier. Check the gas key by feeling for any movement and visually inspect the stake screws that attach it to the carrier body. Inspect for burs on the front of the gas key and an overall visual inspection of the carrier body.

Figure 2.19. Check gas key

g. Bolt. Inspect the gas rings to ensuring that the gaps are NOT lined up. Check the extractor for positive spring tension. Make sure there are no brass shavings underneath the extractor claw. Push in on the ejector checking for positive spring tension. Visually inspect every lug ensuring there are no cracks forming or missing lugs. Check the firing pin protrusion hole for excessive wear or anything that would prevent the firing pin from moving freely.

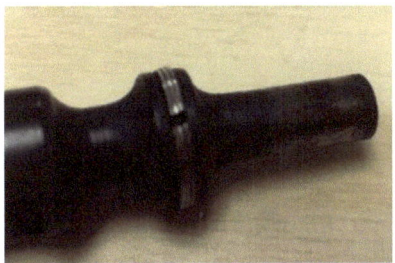

Figure 2.20. Improper gas ring alignment

Figure 2.21. Proper gas ring alignment

h. Bolt cam pin. Inspect where the firing pin passes through for cracks and anything that would prevent the cam pin from moving freely.

i. Firing pin. Check for excessive wear, straightness and any erosion on the tip of the firing pin. Check the firing pin retaining pin for straightness. Remove all carbon and fouling.

j. Charging handle. Check for straightness. Ensure the roll pin for the catch lever is not protruding in either direction.

k. Lower receiver. NOTE – DO NOT DRY FIRE THE WEAPON DISSASEMBLED. Check the safety lever for rotational movement. Check the spring tension on the bolt release and magazine release for positive spring tension. Check the castle nut for tightness and is not loose in any way. Check the trigger and hammer pin holes for excessive wear.

l. Butt stock. Check for cracks and ensure that it locks into every position. Check that the sling swivel screw is tight.

m. Ensure the grip is tight. A loose grip will result in improper function of the selector lever.

n. Upper receiver. Grasp the upper receiver and barrel forward of the front sight assembly, rocking left to right checking to see if there is any movement between the barrel and upper receiver.

Figure 2.22. Check for movement between barrel and upper receiver

o. Check the dust cover spring and forward assist for positive spring tension. Inspect the front sight post for straightness and overall wear. Ensure the gas block retaining pins are seated and not loose. Check the gas tube roll pin is present and not protruding. Check the gas tube for straightness by sliding the stripped bolt carrier within the upper receiver. The bolt carrier should move freely within the upper receiver into the fully seated position. Depress the delta ring and remove ONLY the lower hand guard cover to inspect and clean the barrel.

p. Optics and attachments. Check for tightness, both the mount and the optic or attachment, every time the weapon is drawn from the arms room. Tie downs will be by unit Standard Operating Procedure.

q. Inspect magazine feed lips for cracks and excessive wear. Check the follower moves freely under spring tension.

Figure 2.23. Inspect magazine

2-6. **CLEANING AND LUBRICATION OF THE M4/M16 SERIES RIFLE**

a. Buffer tube spring should be cleaned by inserting a cleaning towel or like item through the spring at one end and rotating the spring until reaching the opposite end. Wipe clean the buffer assembly and buffer tube, inside and out.

b. Remove any foreign debris or fouling from the trigger well and components. Place a thin coat of CLP on all parkerized steel surfaces and clean and dry all anodized aluminum surfaces.

c. Upper receiver. Weapons cleaning should follow the path of the bullet, from the chamber to the muzzle. Saturate the chamber brush with CLP and insert into the chamber, rotating it clockwise. Do not insert so far that the barrel extension bristles go into the chamber, damaging your brush. Push CLP soaked 5.56mm cleaning patch down the barrel with a jag saturating the bore of the weapon with CLP. Push the 5.56mm bore brush down the barrel to remove fouling. Push a dry patch through the barrel to remove all remaining CLP and fouling. These steps are repeated as necessary until the barrel is clean. **Remember to clean from the chamber to muzzle each time, do not drag the brush or patch back and forth in the barrel.** Apply a thin coat of oil to the muzzle device.

Figure 2.24. Clean barrel from breach end

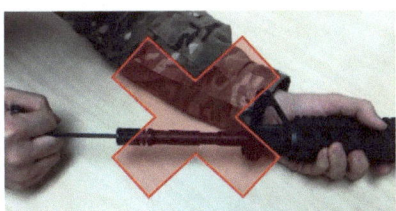

Figure 2.25. Improper barrel cleaning

d. A light coat of oil may be applied down the barrel ONLY for storage. A dry patch must be pushed down the barrel to remove the oil before firing. Barrels on the M4/M16 series of rifles are chrome lined and will not rust if stored for short periods of time dry. The outside of the barrel still requires a light coat of oil to prevent rust.

e. Clean the inside of the upper receiver with a nylon brush and CLP. Wipe away all remaining fouling and excess CLP and wipe dry. Apply a light coat of oil to all parkerized components for rust prevention, i.e. forward assist and dust cover. All exterior components should be cleaned with a nylon brush being careful not to remove the parkerized (steel) or anodized (aluminum) protective finishes. All parkerized component require a light coat of CLP, all anodized components can simply be wiped clean and left dry.

f. Bolt carrier group.

    1. Bolt carrier. Remove all carbon and fouling from the inside and outside of the bolt carrier and inside the gas key. Never insert a Q-tip inside the gas key, only use pipe cleaners.

    2. Apply a light coat of oil to the gas rings and bearing surface of the bolt body.

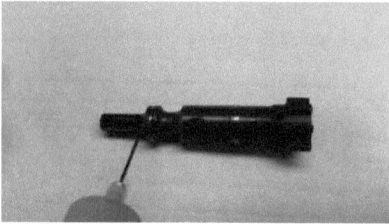
Figure 2.26. Lubricate gas rings

Figure 2.27. Lubricate bolt bearing surface

    3. Apply a thin coat of oil to all parkerized surfaces for rust prevention.

    4. Apply a thin coat of oil to the bolt cam pin, firing pin and firing pin retaining pin.

    5. Apply a light coat of oil to the operational bearing surfaces of the bolt carrier group. These can be identified by the visible wear and removal of parkerizing through friction.

g. Once the weapon is re-assembled, perform a functions check ensuring to listen for the metallic click of trigger reset and ensure the safety lever functions properly.

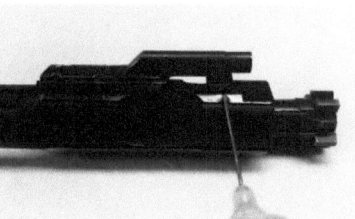
Figure 2.28. Lubricate bolt carrier group bearing surface

CHAPTER 3 - FUNDAMENTALS OF MARKSMANSHIP

GENERAL

3-1.   In order to accurately engage a target, the Soldier essentially has to perform primary two tasks:

a. Point the rifle at the target (sight alignment).

b. Fire the rifle without moving it (trigger control).

These two steps are the two fundamentals of shooting. They apply to all shooting, regardless of the event or type of rifle used. As you practice and study shooting, you will notice there are many techniques used successfully to accomplish a given task. The fundamentals, however, are the same for everyone, every time.

FUNDAMENTALS OF RIFLE MARKSMANSHIP

3-2.   The two fundamentals of rifle marksmanship are Aiming (Sight Alignment/Sight Picture) and trigger control also known as the integrated act of firing. In order to be a proficient rifleman, you have to be able to master all of the fundamentals of marksmanship. By becoming proficient in these skills, you will multiply your value on the battlefield. Each fundamental will be discussed in detail in the following paragraphs.

AIMING

3-3.   Aiming is the process a rifleman uses to point the rifle at the target. Aiming is comprised of sight picture and sight alignment.

SIGHT ALIGNMENT

3-4.   Sight Alignment is defined as centering the tip of the front sight post in the center of the rear sight aperture both vertically and horizontally, and aligning the relationship with the shooter's dominant eye. The following illustrations depict the concept of sight alignment.

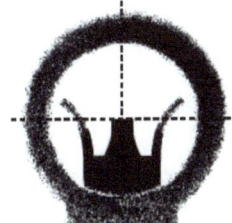

Figure 3.1. Imagine a vertical line drawn through the center of the rear sight aperture. The line will appear to bisect the front sight post.

Figure. 3.2. Imagine a horizontal line drawn through the center of the rear sight aperture. The top of the front sight post will appear to touch this line.

3-5.   Consistent Sight Alignment is achieved by resting the full weight of your head on the stock in a manner that allows your dominant eye to look through the rear sight aperture. This is also referred to as stock weld. It is NOT necessary to put your nose on the charging handle to get good stock weld.

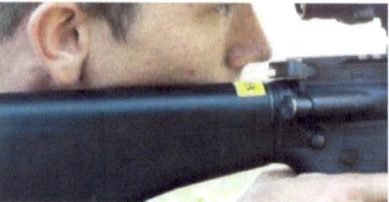

Figure 3.3. Improper stock weld.

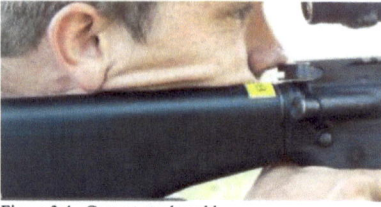

Figure 3.4. Correct stock weld.

3-6.   Importance of correct sight alignment. A sight alignment error results in a misplaced shot through angular error. The error grows proportionately greater as the distance to the target increases. An error in sight picture, however, will remain constant regardless of the distance to the target. The figure on the right depicts the impact of misaligned sights when the weapon is fired. In Figure 3.5, the illustration on the left side of the figure depicts the front sight grossly misaligned within the rear sight, to the right of center. The effect of this on the bullet path is shown on the right side of the figure and is going to cause the bullet to be misplaced to the right. At 100 yards this error may still allow you to hit your intended target, however at 600 yards this error could cause you to miss your intended target.

How small of an error can effect accuracy? For every one thousandth of an inch of error the bullet will be displaced one inch at 300 yards. For your point of reference a human hair or the thickness of a piece of notebook paper is 4-5 thousandths of an inch.

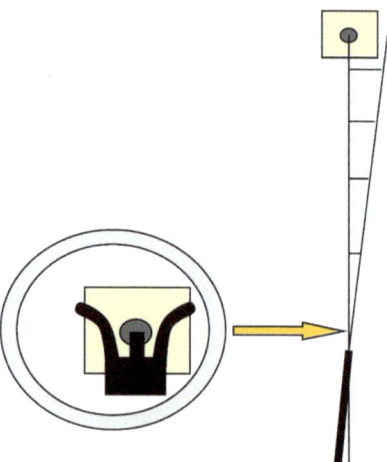

Figure 3.5. Effect of improper sight alignment

If you concentrate on the relationship of the front sight and the rear aperture, you will be able to hold to a high level of accuracy.

SIGHT PICTURE

3-7.   Sight picture is the placement of the tip of the front sight post in relation to the target while maintaining sight alignment. Correct sight alignment but improper sight placement on the target will cause the bullet to impact the target incorrectly on the spot where the sights were aimed when the bullet exited the muzzle.

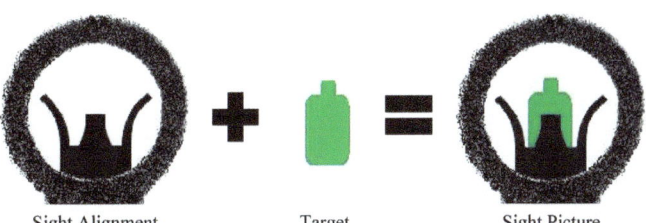

Sight Alignment   Target   Sight Picture

Figure 3.6. Definition of sight picture

EYE DOMINANCE

3-8.   Most Shooters have a dominant eye, one that is stronger than the other. In order for you to aim precisely, you should use your dominant eye. An easy way to determine your dominant eye is to hold your hands out with the fingers extended and joined, thumbs extended out to the sides. Overlap your hands at a 90-degree angle until a small window is made with your thumbs. Place your hands out at arm's length and look through the opening in your thumbs at an object (Figure 3-7). Without squinting or closing either eye, bring both hands to your face while maintaining visual contact with the object. The hole will move to your dominant eye as your hands reach your face. If you are still not sure, simply have another individual stand back from you at least 15ft and look at their face through the hole in your hands. They will be able to see your dominant eye through the hole. It is not uncommon for about 20% of any group to find that they are cross dominant. Many riflemen are successful shooting right or left handed in order to aim with their dominant eye. Although it may seem awkward at first, it is possible to change your firing hand in order to use your dominant eye.

Figure 3.7. Determining the dominant eye

POINT OF FOCUS

3-9.   The eye can only focus on one object at a time. Given two sights and a target, the Shooter has several things to look at while aiming. The best technique is to focus on the tip of the front sight or optic reticle while maintaining sight alignment. This technique provides riflemen with the most consistent results regardless of conditions. While the process of aiming requires the Soldier to shift focus to identify the target and achieve a good sight picture and sight alignment, the final focus at the instant the rifle fires should be on the tip of the front sight post ensuring that you have maintained sight alignment. Most new Soldiers have a tendency to focus on the target since that is what they are trying to hit. This technique results in an out of focus front sight, misalignment of the sights, and inconsistent aiming. With practice, any Soldier will notice dramatically better accuracy when their focus is concentrated on the front sight or optic reticle, and sights that are aligned. The pictures below depict correct and incorrect focus.

CORRECT

Figure 3.8. This is an actual photo of correct sight alignment with issued iron sights. Take note of how blurry the target is in relation to the front sight post. Unless the front sight post is THIS clear, sight alignment is questionable and exactly where your rifle is pointed is not certain.

INCORRECT

Figure 3.9. This photo illustrates a clear target. Note that the front sight post is lost within the target. If a Soldier were to fire with this lack of focus, inaccuracy or an outright miss is guaranteed.

TRIGGER CONTROL

3-10. Now that you understand how to properly point the rifle at the target, the next task is to fire the rifle without disturbing sight alignment. This process is called trigger control because that is exactly what the Soldier is doing, controlling the trigger. Trigger control is the skillful manipulation of the trigger straight to the rear that causes the rifle to fire without disturbing sight alignment or sight picture. When shooting, the rifleman controls where the rifle is pointed and when it fires by deliberate movement of the trigger. Manipulating the trigger requires a mental and physical process. Controlling the trigger is a mental process, while pulling the trigger straight to the rear is a physical process. There are two types of trigger control:

a. Uninterrupted trigger control. The preferred method of trigger control in a combat environment is uninterrupted trigger control. After obtaining sight picture, the Shooter applies smooth, continuous pressure rearward on the trigger until the shot is fired.

b. Interrupted trigger control. Interrupted trigger control is used at any time the sight alignment is interrupted or the target is temporarily obscured. An example of this is extremely windy conditions when the weapon will not settle, forcing the Soldier to pause until the sights return to his aiming point. To perform interrupted trigger control: move the trigger to the rear until an error is detected in the aiming process. When this occurs, stop the rearward motion on the trigger, but maintain the pressure on the trigger, until the sight picture is achieved. When the sight picture settles, continue the rearward motion on the trigger until the shot is fired.

3-11. Resetting the Trigger. During recovery, release the pressure on the trigger slightly to reset the trigger after the first shot is fired (indicated by an audible click). Do not remove the finger from the trigger. This places the trigger in position to fire the next shot without having to reestablish trigger finger placement.

3-12. A firm grip is essential for effective trigger control. The grip is established before starting the application of trigger control and it is maintained through the duration of the shot. To establish a firm grip on the rifle, position the "V" formed between the thumb and index finger on the pistol grip behind the trigger. The fingers and the thumb are placed around the pistol grip in a location that allows the trigger finger to be placed naturally on the trigger and the thumb in a position to operate the safety. Once established, the grip should be firm enough to allow manipulation of the trigger straight to the rear without disturbing the sights.

Figure 3.10. Firm grip

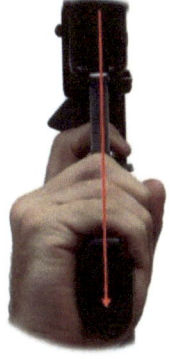
Figure 3.11. Pull trigger straight to the rear

3-13. Correct trigger finger placement allows the trigger to be pulled straight to the rear without disturbing sight alignment. The trigger finger should contact the trigger naturally. The trigger finger should not contact the rifle receiver or trigger guard. The place where the index finger contacts the trigger is individually based. Soldiers with short fingers will have the first meaty portion of their finger on the trigger, while Soldiers with long fingers may have up to the second joint of the index finger in contact with the trigger. A proper hand position is achieved when the wrist is not bent at an unusual angle, but remains in-line with the firing arm.

Figure 3.12. Incorrect trigger finger placement

3-14. Taking up the "slack" or applying initial pressure to the trigger is also an integral part of trigger control. This means that you apply some pressure to the trigger from the start, taking up the mechanical slack or creep in the trigger. Improper trigger control will disturb the rifle during firing. Begin applying initial pressure as you acquire your target, and as your final focus goes to your front sight you should be causing the weapon to fire by pulling the trigger smoothly and straight to the rear.

FOLLOW THROUGH AND RECOVERY

3-15.   The follow through process is nothing more than the continued application of the fundamentals until the round has exited the barrel. In combat, follow through is important because it completes the action and ensures proper form allowing for consistency. Follow through is continuing to aim and apply pressure to the trigger after the rifle fires. Follow through prevents the Soldier from anticipating the shot and movement prior to the bullet leaving the barrel. How long the Soldier must follow through may vary from one to another. One to three seconds (about the time it takes to recover from recoil) is typical.

3-16.   It is important to get the rifle sights back on the target for another shot. This is known as recovery. Shot recovery starts immediately after the round leaves the barrel. To recover quickly, a Soldier must physically bring the sights back on target as quickly as possible.

CHAPTER 4 - STEADY POSITION FUNDAMENTALS

GENERAL

4-1.   Without a consistent, steady position your ability to apply the fundamentals will be limited. There is no "cookie cutter," method for putting Soldiers into proper shooting positions, because each Soldier must be able to build a fundamentally correct position. There may be variations from the way one Soldier's position looks compared to another. Each Soldier is different in body shape, size, and flexibility; but each position must enhance and support aiming, trigger control and follow up shots. In this section you will learn the key elements common to all good positions and how they can be properly used in prone, sitting, kneeling, and standing.

HEAD POSITION AND SIGHT ALIGNMENT

4-2.   As mentioned above, every shooter has a different body shape, size, and flexibility, some characteristics of building a position may change. For your position to support proper sight alignment you must place your head on the stock the same way for every shot. Your dominant eye should be directly in line with an imaginary plane that runs from the top center of the front sight post through the center of the rear sight aperture. It sounds complicated, but is actually very simple. As you learned previously your eye can find the center of the rear sight (a circle) naturally.

Head position

Figure 4.1.   Proper head position

a. First bring the rifle up to your head and let your head fall naturally onto the rifle stock. Your head should be generally erect and level if you were to look at it from either the front or the side.

b. Correct vertical placement of the butt stock in the shoulder will allow the rifle to be brought up to your face in a more natural, upright manner. Do not drop your head down to the rifle. Starting with the bottom of the jawbone, drag the firing side of your face in a downward movement across the top of the stock until you are resting the full weight of your head on top of the stock. The side of your face should be in firm contact with the rifle stock.

Figure 4.2.   Proper head position from the front

c. In all positions, your dominant eye should be in line with an imaginary line that runs from the tip of the front sight post directly through the center of the rear aperture. If you look through your sights and notice that you are not looking directly through the rear aperture at the front sight post, your head position on the stock is incorrect. A comfortable, relaxed position is one that can be used effectively. This will only be possible if your head is completely supported by the rifle stock.

d.

e. While live firing it is important that your head and the rifle move as a single unit during recoil. To ensure the head and rifle recoil together, the rifle should be firmly placed or pulled into the firing shoulder (not necessarily the pocket), with your head completely supported by the stock. Simply put, the application of sight alignment is correct head position on the stock.

EYE RELIEF

4-3.   Eye relief is the distance between your eye and the rear sight aperture. Eye relief varies from Soldier to Soldier and between positions. The rifleman must practice <u>consistent</u> eye relief for a given position. This is achieved by consistent stock weld. Ideally, your sights should be straight up and down. Canting or tilting the rifle will affect your zero. It is much easier to obtain consistent sight alignment with your sights straight up and down.

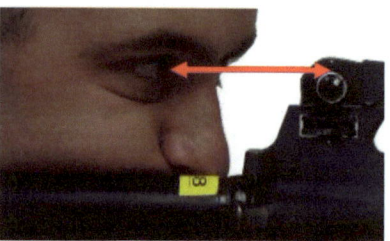

Figure 4.3. Eye relief. Note nose NOT on charging handle

CHARACTERISTICS OF A STEADY POSITION

4-4.   The purpose of a firing position is to support the two basic fundamentals of marksmanship, Sight Alignment & Trigger Control. There are three characteristics common to any good position:

a. **Support (bone and artificial).** Bone support is nothing more than properly using the skeletal frame of your body to transmit the stability of the ground into your position. Artificial support is anything external to your own body that you may use to stabilize your position (sling. etc.). Maximizing bone and sling support reduces muscle tension, which is the second characteristic.

b. **Muscular Relaxation.** When you are in a position that uses good bone or artificial support, your muscles will tend to be more relaxed. This is also very important in getting your position to have a slow, small wobble area. In general, a position that has less muscle tension has less movement.

c. **Natural Point of Aim (NPA).** After establishing good support your muscles may be more relaxed, and your position will be pointing somewhere naturally. This is your natural point of aim. To determine exactly where your NPA is pointing:

1. Relax and exhale while closing your eyes. Open your eye(s) and look through your sights this is your natural point of aim.

2. This will be adjusted by shifting your position so that you are aligned naturally on the target. This must be adjusted vertically and horizontally within your position.

ELEMENTS OF A STEADY POSITION

4-5.   There are five elements to a steady position that we will focus on:

a.   **Stock Weld.** Proper stock weld is accomplished by keeping your head in the most vertical position possible, while resting the full weight of your head on the stock of the rifle with your dominant eye looking through the sights.

b.   **Rifle butt.** Place the butt of the rifle high enough in your shoulder to bring the rifle up to your face, allowing your head to rest on the stock in a more natural manner. In the prone, the proper placement will generally be in the pocket of the shoulder.

c.   **Firing-hand.** The firing-hand grasps the pistol grip to allow for proper trigger control. The hand naturally grasps the rifle high up on the pistol grip. Your handgrip should be similar to a firm handshake; this will prevent your whole hand from tensing and allow your trigger finger to move independently to the rear.

d.   **Trigger finger.** Natural placement of the trigger finger on the trigger. (See chapter 3.)

e.   **Breath control.** Breathing causes movement of the chest and will in turn move the rifle sights in relation to the target. Although the breath can be held at any point in the respiratory cycle, it is strongly recommended to hold your breath after a normal exhalation during what is known as the normal respiratory pause. Ensure that you are not holding your breath for an extended period of time, as your vision is the first thing to decline. *Note:* **While a popular topic of discussion, breath control remains secondary to sight alignment and trigger control.**

PRONE POSITION

4-6.   Prone is the most stable position as a large portion of the body is in contact with the ground. The following outlines the steps to take in order to obtain the best prone position.

a.   Assuming the Prone Position.

1.   Point – Point the rifle down range towards the target.

Figure 4.4.   Point

2. Post – Post your non firing hand on the ground naturally in front of you.

Figure 4.5. Post

Figure 4.6. Sprawl

3. Sprawl – Kick your legs rearward and out so that your body lays naturally on the ground in a good firing position.

4. There are two methods for positioning your legs while in the prone position. The first is the bent knee position. The leg on the firing side is drawn slightly towards the shoulders to allow the diaphragm to rise off the ground. The second position is the straight leg position. Both legs are extended straight to the rear creating the greatest amount of contact with the ground. Either method presents a stable firing position and can be used interchangeably at the shooter's discretion.

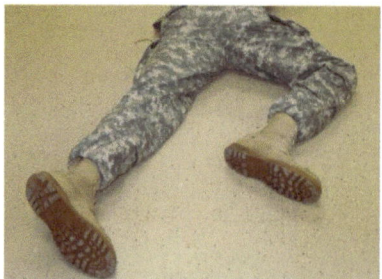

Figure 4.7. Bent knee prone position

Figure 4.8. Straight leg prone position

b. Check your natural point of aim. To do this you must close your eyes and relax, take a few deep breaths and open your eyes. Take note of where your position is pointing. If you are not naturally aligned with the target you must readjust your position using the following technique:

1. To move the rifle to the right, shift your hips to the left while keeping your non-firing elbow in place. This elbow will act as a pivot point for all adjustments.

2. To move the rifle to the left, shift your hips to the right.

3. To depress the muzzle, shift your hips forward.

4. To elevate the muzzle, shift your hips to the rear.

You will notice that very small movements are necessary to refine your natural point of aim, especially at longer ranges. Check your position continually throughout your adjustments and while firing. Ensure that you have ready access to your ammunition. Be consistent and meticulous in your habits and the shot-making process.

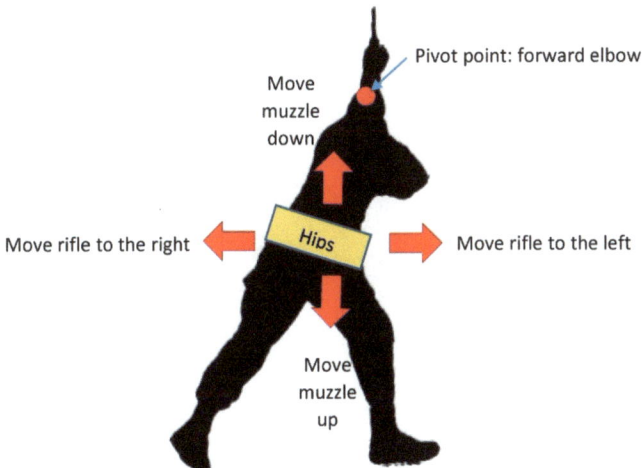

Figure 4.9. Adjusting prone natural point of aim

SITTING POSITION

4-8.   Sitting is the second most stable platform after prone. The position should afford maximum stability and facilitate sight alignment through an upright and level head position. There are two common variations of the sitting position: the crossed leg and the open legged. Each will be discussed, but only experimentation and dry firing will allow you to determine which is right for you. In both types of positions the head should be fairly erect. You should look through the rear sight in a manner where the neck and eye are not strained. The rifle butt should be placed in the shoulder high enough to keep the head and neck erect. Each step taken should complement a proper stock weld.

a. Crossed Leg Position:
Bend forward at the waste while placing your support elbow on your support leg into the pocket of the knee. Lower your firing elbow into the inside of the firing side knee. Check your natural point of aim. If you discover that you need to adjust your natural point of aim, the only part of the position that moves is the buttock. Sliding to the left or right will adjust side to side. Moving closer or further from the legs/ankles will change to elevation. You can also move your non-firing hand in or out to make a small change to elevation.

Figure 4.10.   Crossed leg sitting position

Figure 4.11.   Open leg sitting position

b. Open Leg Position:
The Open Legged Sitting Position is the highest profile Sitting Position. To assume the position, begin standing with feet approximately shoulder width apart and face approximately at a 45 degree angle to your target. While dropping to a knee, sit down and extend legs into an open position. Bend forward at the waste while placing your support elbow on your support leg into the inside of the leg while avoiding bone to bone contact. Likewise, lower your firing elbow into the inside of the firing side leg. NOTE: This position may require controlled muscular tension to maintain.

STANDING POSITION

4-11.   The standing position is the least stable position, but if you understand a few simple concepts to the position, you will have a more successful time shooting standing. A more stable position will produce a better hold. The more balanced you are, the better your hold.

Figure 4.12.   Standing position

a.   Begin with your feet slightly more than shoulder width apart in an athletic stance with your chest and shoulders squared with your target and your feet offset with your non-firing foot slightly forward of the other. To achieve sustainable balance, imagine keeping your chin in front of your hips in a slightly forward lean.

b.   For placement of your non-firing hand, it should be as far forward on the hand guard as possible while maintaining a slight bend in the elbow to absorb recoil and transition the rifle from target to target. Once achieving this position, maintain a *slight* rearward pressure to provide additional recoil management.

Figure 4.13.   Standing head position

c.   Your firing arm should be at where it falls naturally. Don't lift your elbow up or concentrate on keeping it pressed down. Your firing hand is placed firmly on the pistol grip. As with the other positions, the trigger finger needs to be placed to be able to pull straight to the rear without disturbing the weapon.

Figure 4.14.   Standing firing arm placement

d.   Place the butt stock into your shoulder where you can achieve an optimum head position. Head position is paramount in the standing position and must be upright and level free of cant. To achieve this, bring the stock up to your face and look through the sights.

KNEELING POSITION

4-12.   The kneeling position is not as stable as the prone or sitting positions, but allows the Soldier to move in and out of a relatively stable shooting position quickly.

Figure 4.15.   Kneeling position

a.   To assume the kneeling position, take one step forward with your non-firing side foot and drop your firing side knee to the ground. Sit back on your firing foot creating three points of contact with your center of gravity over your firing foot.

b.   Turn your non-firing foot 45 degrees inwards towards your body creating tendon lock out in your non-firing ankle and calf.

c.   Place your non-firing elbow comfortably on your non-firing leg without creating bone-to-bone contact. There are two common placements for this, either with the elbow above the knee on the thigh or the triceps muscle resting forward of the knee.

d.   Your firing arm should be at where it falls naturally. Don't lift your elbow up or concentrate on keeping it pressed down. Your firing hand is placed firmly on the pistol grip. As with the other positions, the trigger finger needs to be situated to achieve the best mechanical advantage.

Figure 4.16.   Kneeling firing arm placement

Figure 4.17.   Kneeling non-firing arm placement

e.   Place the butt stock into your shoulder where you can achieve an optimum head position. Head position is paramount in the standing position and must be upright and level free of cant. To achieve this, bring the stock up to your face and look through the sights.

f. An alternate kneeling position is achieved by coming off of your firing foot and shifting your center of gravity over your non-firing foot. This allows you to move in and out of position quicker and is more aggressive in nature.

Figure 4.18. Alternate kneeling position

CHAPTER 5 - SHOT PROCESS

5-1   The shot process is a detailed list of tasks in a specific order that the Soldier should execute for each shot taken.

1.	PRE SHOT	Assume position
2.		Natural point of aim
3.		Sight alignment / sight picture
4.		Point of aim
5.	SHOT	Refine sight alignment / sight picture
6.		Apply trigger control
7.		Follow through
8.	POST SHOT	Call the shot
9.		Recoil recovery
10.		Evaluate

Table 5.1. Shot process

5-2.   PRE SHOT

a. Assume Position - the firer will assume the position that gives them the most stability or the position designated orienting their weapon to the target

b. Natural Point of Aim (NPA) - this will be achieved by adjusting the assumed firing position so that the sights will be aligned with the target with a minimal amount of muscle tension. NPA can be checked by closing your eyes, relaxing the position, reopening your eyes, and checking to see if the rifle has remained on target.

c. Sight Alignment/Sight Picture - as stated in Chapter 3 this is the relationship between the sighting system and the firer's eye. The process used by a Soldier depends on the sighting system employed with the weapon. Iron sight/back up iron sight. The firer aligns the tip of the front sight post in the center of the rear aperture and his/or her eye. The firer will maintain focus on the front sight post, simultaneously centering it in the rear aperture.

d. Point of Aim (POA) - is the location on the target on which the firer places the tip of the front sight/or the reticle of the optic. This will change based on the zero of the rifle, range of the target, or target exposure. Wind will also be a factor for POA this will be discussed in further detail in chapter 9.

5-3.   SHOT (Note: the steps to firing the shot, refining sight alignment/sight picture and trigger control must be accomplished simultaneously.)

a. Refine Sight Alignment/Sight Picture- is refined by seeing the *very* tip of the front sight clearly centered in the rear aperture or clearly seeing the reticle of the optic. This is placed on the point of aim while the trigger is squeezed.

b. Apply Trigger Control - as stated in Chapter 3 the soldier will squeeze the trigger in a smooth consistent manner straight to the rear- consistently adding pressure until the weapon fires. The soldier will hold the trigger to the rear until the recoil is complete, and then reset the trigger without removing their finger from the trigger. Regardless of the speed at which the soldier is firing trigger control will always be smooth.

c. Follow Through- as stated in Chapter 3 the continued mental and physical application of the firing tasks after the round has been fired. This includes resetting the trigger, reacquiring the sights, and returning the sights back to the original hold. Follow through actually bridges the shot process between SHOT and POST SHOT.

5-4. **POST SHOT**

a. Recoil Recovery - after the shot is fired the firer will follow the rifle back to the target and attempt to recover his natural point of aim and sight picture.

b. Calling Shot - as stated in Chapter 3 this refers to the location of the sights at the time the shot breaks. More specifically this refers to a firer stating exactly where he thinks a single shot has hit by looking at the sights relationship to the target when the rifle fires. This is normally expressed in clock direction and inches from the desired point of aim; i.e. 9 o'clock and 3 inches.

c. Evaluate - it is a summary of the entire shot process to include down range feedback to make adjustments for additional follow on shots.

CHAPTER 6 - BALLISTICS

6-1.   Ballistics is everything that happens to the round from the time the firing pin strikes the primer to the time that the bullet enters the target and comes to a complete stop. There are three types of ballistics: Internal, External, and Terminal.

   a.   Internal Ballistics concerns what happens to the projectile before it leaves the muzzle of the rifle.

   b.   External Ballistics deals with factors affecting the flight path of the projectile between the muzzle of the rifle and the target. For marksmanship purposes only external ballistics will be covered in this book.

   c.   Terminal Ballistics deals with what happens to the projectile when it comes in contact with the target.

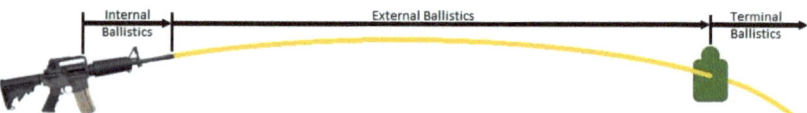

Figure 6.1. Internal, external, terminal ballistics

6-2.   There are two factors that affect the path of a bullet: Gravity and air resistance.

   a.   Gravity – Immediately upon exiting the muzzle, gravity causes the bullet to fall towards the earth. The way to counter gravity is to increase the angle of departure by elevating the muzzle.

   b.   Air resistance (Drag) – Causes the bullet to slow down, fly erratically, and eventually destabilize and tumble. The way to counter the forces are to increase muzzle velocity and impart spin on the bullet.

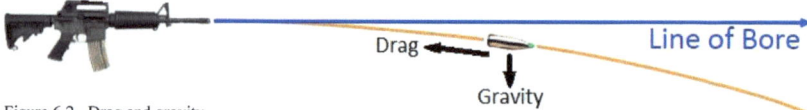

Figure 6.2. Drag and gravity

6-3.   Effects of Weather. Other external factors that influence trajectory relative to the point of aim are wind, temperature, altitude, humidity, and barometric pressure. Wind is the most significant of these factors.

6-4.   Wind affects the trajectory of the projectile. The longer a projectile is exposed to wind the greater the affect it has on the projectile down range. See figure 6.3.

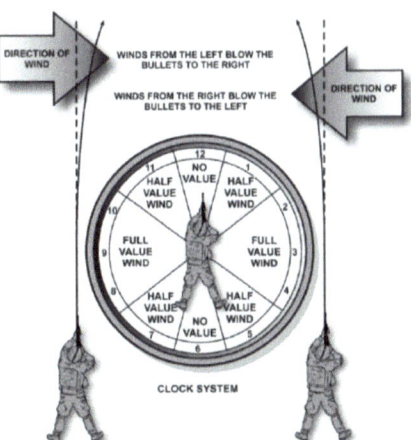

Figure 6.3. Wind chart

6-5.   Trajectory. When a projectile exits the muzzle, gravity immediately takes affect and pulls the bullet below the line of bore toward the ground, creating an arc. As the bullet travels down range the velocity also decreases due to air resistance (drag). The slower the bullet goes, the more time gravity has to work on it. This causes the bullet to depart from the line of bore faster at longer distances making the trajectory a parabolic curve. As distance increases the angle of the line of bore must be increased to ensure proper shot placement.

a.   Line of Sight – What the shooter sees when looking through the sights at the target. This can be illustrated by drawing an imaginary line from your eye through the sights to the target.

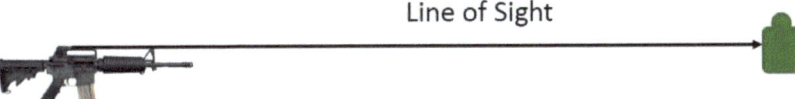

Figure 6.4. Line of sight

b.   Line of Bore – An imaginary line drawn through the center of the bore to infinity.

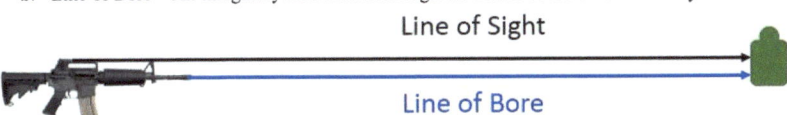

Figure 6.5. Line of bore

c.   Trajectory – The path that the bullet will take when it is fired from the rifle, as affected by gravity and drag. Also affected by wind.

Figure 6.6. Trajectory (path of bullet)

6-6.   Minute of Angle (MOA)

a. MOA is the standard unit of angular measurement used in adjusting rifle sighting devices and other ballistic related measurements. It is also used to define the accuracy of a rifle.

b. MOA is an angular unit of measure (1/60th of a degree) which very nearly equates to 1 inch per 100 yards (2 inches at 200 yards, 3 inches at 300 yards, etc.) MOA can be used interchangeably for yards and meters at distances closer than 300 meters because the difference between them is negligible. Smaller measurements are described in fractions. (i.e. ½ MOA).

c. The new Army M4/M16 25m Zero Target is calibrated in MOA (Chapter 7, Figure 7.13).

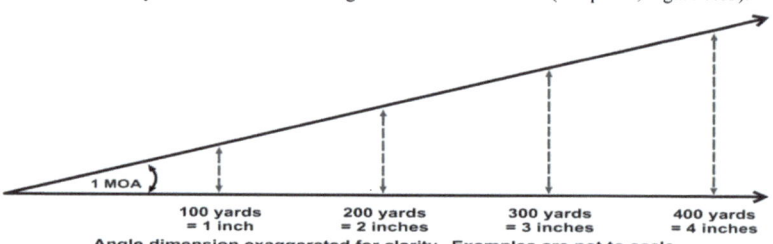

Figure 6.7. Minute of angle

CHAPTER 7 - ZEROING

GENERAL

7-1.   Zeroing adjusts the sights to center the group on the target at a given distance. A zero is unique to a specific rifle, sight, and shooter combination. Changing the Soldier firing a rifle may change the zero.

7-2.   There are two common types of zeroes used with the M16 and M4 rifles.

a. **Battle sight zero** – The elevation and windage settings required to effectively engage targets from 0-300 meters given standard issue ammunition with minimal hold-offs (figure 7.1). This is the standard zero used with the M16 or M4 zero targets at 25 meters. For M16 and M4 rifles with iron sights adjust the sights so that the point of aim and point of impact correlate at 25 meters. Confirm at 300 meters. The windage zero should be for a no-wind condition at distance.

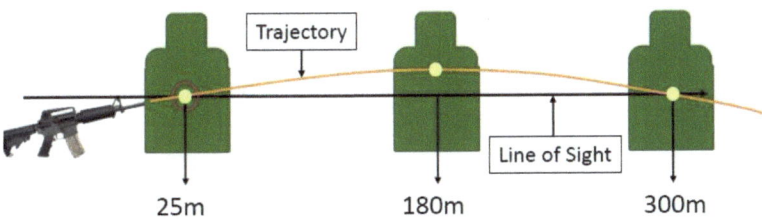

Figure 7.1. Battle sight zero

b. **Bullet drop compensating zero (BDC)** – Initially zero the rifle the same as the battle sight zero. The BDC allows you to engage targets at known distances past 300 meters by using the distance indicators on the back up iron sight or the carry handle sight without holding over the target. The BDC elevates the sights a predetermined amount to allow for a center of mass hold at extended ranges. To work correctly, it is imperative to properly zero the rifle at 300 meters.

7-3.   **Grouping and zeroing procedures (25/300 meters)**

a. The goal of the grouping exercise is to fire shot groups and consistently place those groups in the same location. Tight shot groups demonstrate that the shooter is applying the proper fundamentals before beginning the zeroing process. It is not possible to achieve an accurate zero if the shooter cannot fire tight groups.

b. PMCS equipment according to TM prior to conducting any operations.

c. With an <u>unissued</u> weapon, mechanically zero the front and rear sights.

2. To mechanically zero the front sight adjust it up or down until the base of the front sight post is flush with the top of the front sight housing as shown in figure 7.2.

Figure 7.2. Front sight mechanical zero

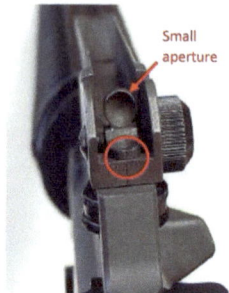

Figure 7.3. Rear sight mechanical zero

3.   Ensure that the smaller of the rear sight apertures is up if using a carry handle sight, and visually ensure that the line on the rear sight aperture is aligned with the center line on the windage scale. Lower the elevation wheel to the lowest position, figures 7.4-7.6. With the backup iron sight (BUIS) ensure the windage index mark is centered on the windage scale, figures 7.7 & 7.8.

d. When zeroing at 25 meters use the rear sight settings in Table 7-1.

RIFLE		SIGHT SETTING
M16A2/A3	(Figure 7.4)	8/3+1 click up
M16A4	(Figure 7.5)	6/3+2 clicks up
M4	(Figure 7.6)	6/3
M16 BUIS	(Figure 7.7)	Set to zero mark between 300 and 400
M4 BUIS	(Figure 7.8)	Set to 300

Table 7.1. Rear sight zero settings

Figure 7.4. M16A2/A3 zero setting   Figure 7.5. M16A4 zero setting   Figure 7.6. M4 zero setting

Figure 7.7. M16 BUIS zero setting

Figure 7.8. M4 BUIS zero setting

e. When zeroing at 25 meters on the M16/M4 zeroing target, fire a 5 shot group of well-aimed center of mass shots (figure 7.9) from a supported firing position. Adjust shot placement to get groups centered in the lower half of the 4 cm zeroing circle, which gives a greater probability of hits from 150-250m (figure 7.10). Make elevation changes on the front sight (up and down). Make windage changes on the rear sight (left and right). Carefully read the direction arrows on the sights. UP and RIGHT are labeled.

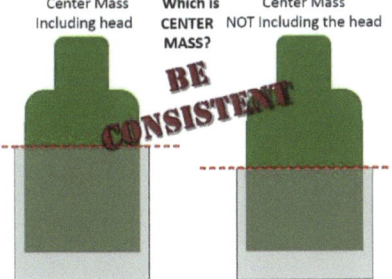

Figure 7.9. Center of mass

f. Confirm zero at 300 meters by shooting 5 round groups, adjusting if necessary to move groups to center mass on an E type target. After 300 meter confirmation is complete, shoot groups at 200 and 100 meters to confirm hold-offs at distance. When confirming at distance, place the sights on 8/3 or 6/3 for your respective weapons system. The extra one or two clicks are not necessary when firing at distance. If the BUIS is used, set it to 300 meters.

7-4.   Conduct fundamental marksmanship training, grouping and zeroing, without combat equipment. Once fundamental marksmanship training is complete, equipment should be worn when conducting combat focused training.

Figure 7.10. Zero in lower half of 4cm circle

7-5.   Your rifle zero is not set in stone. As your skill level improves, or conditions change, your zero may also change. Don't be afraid to adjust your sights as needed. Reconfirm zeroes at every available opportunity.

FACTORS AFFECTING ZERO

7-6.   Several factors listed below may affect the zero of a rifle. Soldiers should be aware of these and reconfirm whenever there is an opportunity.

a. Maintenance. A properly maintained rifle will perform better. Inspect and clean the rifle as described in Chapter 2.

b. Extreme changes in altitude, temperature, humidity. Changes in climate impact the trajectory of a projectile. This will change the zero, especially as range increases. Reconfirm if the opportunity exists.

c. Inconsistent ammunition types or lots. Changing ammunition types or lots may impact the zero.

d. Battle uniform/gear. The extra bulk of equipment will often change a Soldier's positions. This my impact their zeroes.

e. Wind. Wind can play a significant role in zeroing, especially when confirming at distance. Chapter 8 has more details on shooting in the wind and its effects on a bullet.

Figure 7.11. Factors affecting zero

TARGETS

7-7.   Zeroing targets simplify the zeroing process. Figure 7.12 depicts the M16/M4 zero target. The gridlines on this target match the elevation and windage correction necessary on sights, M16 on one side and M4 on the other. Move the sights on the rifle the number of clicks specified by the numbers on the boarder of the target. <u>These targets are specific to a particular weapon system.</u> Ensure the target used matches the weapon and sights.

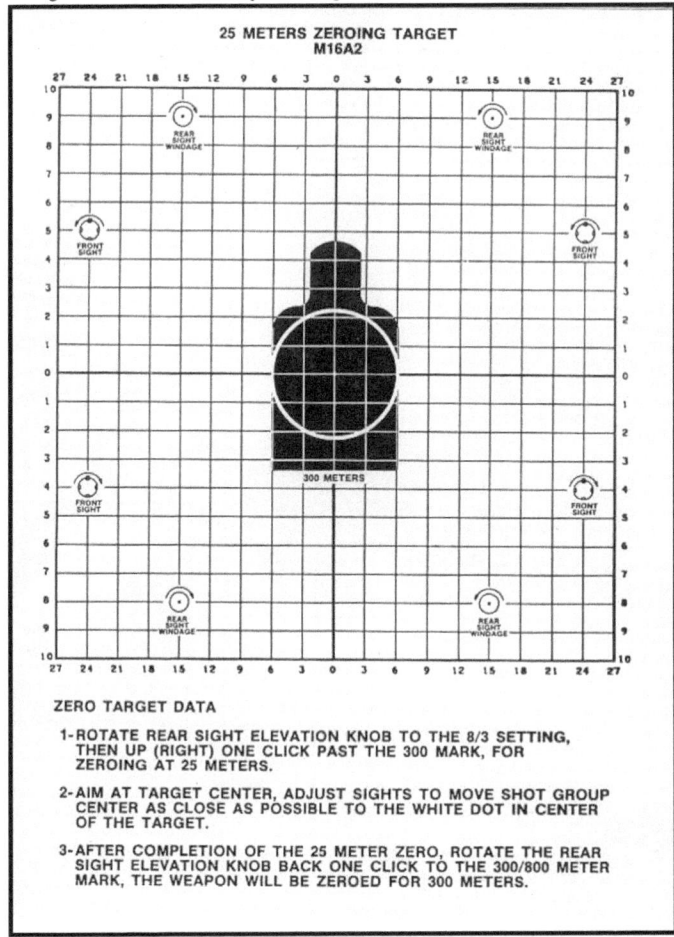

Figure 7.12. M16/M4 zero target

7-8.   Figure 7.13 depicts a new zero target. The gridlines on this target are equivalent to 1 minute of angle at 25 meters. Because of the wide variety of sights used with the M16 / M4 this target offers considerable utility because it will work with all of them. The table at the bottom of the target specifies how many clicks to move the sights to adjust the point of impact by one gridline.

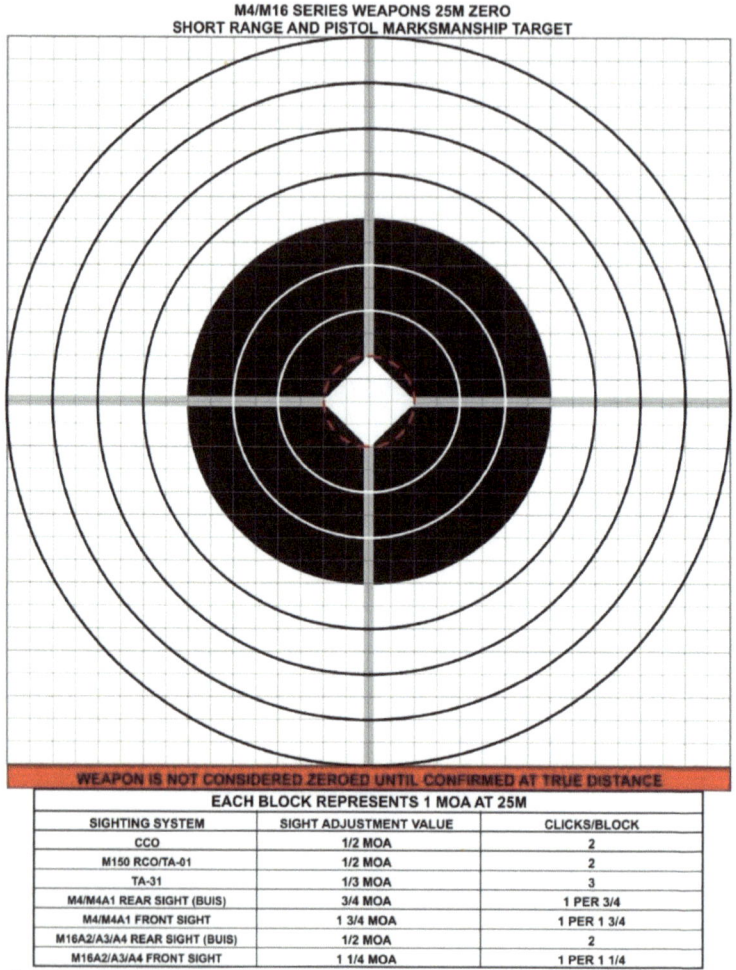

Figure 7.13. New zero target

CHAPTER 8 – TARGET ANALYSIS

8-1.   As a shooter or coach it is important to be able to observe and diagnose shooter errors.

   a.   Shooting should be learned in a safe and relaxed environment, nobody learns a fine motor skill by being screamed at.

   b.   Diagnosing shooter errors will require a dialog between shooter and coach. The only person that can see through the shooters sights during their shot is the shooter. You must ask the shooter what they saw while firing.

   c.   A technique that can be helpful is to use a smart phone and take video of a shooter's firing error, this will allow the shooter the opportunity to see themselves make the error instead of just being told they are wrong.

8-2.   5 shot groups are recommended

   a.   Makes triangulation more accurate.

   b.   Allows for more appropriate sight adjustments.

   c.   Allows shooter to make a minor error on one shot without ruining his group.

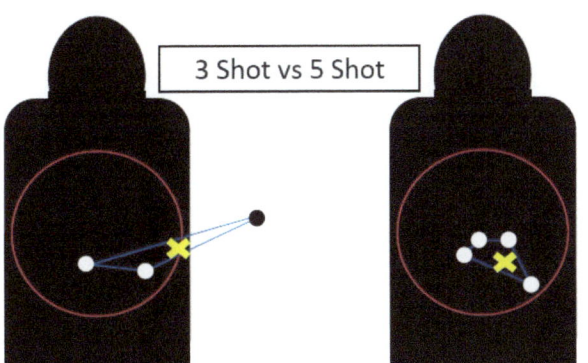

Figure 8.1. 3-shot vs. 5-shot grouping

8-3.   Grouping.

a.   An acceptable 25m group size is 4 out of 5 rounds inside a 2.5cm (1 in) area.

b.   Grouping must show consistency (multiple shot groups in the same 2.5 – 4 cm area before adjustments are made. Not placing group on top of group shows inconsistency in fundamentals.

Acceptable grouping   Unacceptable grouping

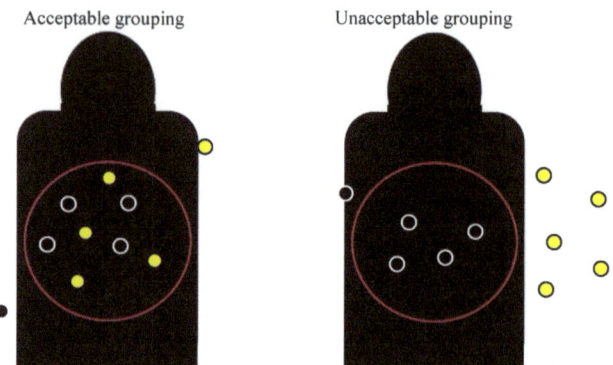

Figure 8.2. Acceptable and unacceptable grouping

8-4.   Shooter improvement.

a.   Failure to shoot a group within the weapon system's capabilities will cause a failure to zero.

b.   As our shooting skills improve and our groups shrink, it is not uncommon for zeroes to shift.

Poor shot grouping   Shot grouping improvement

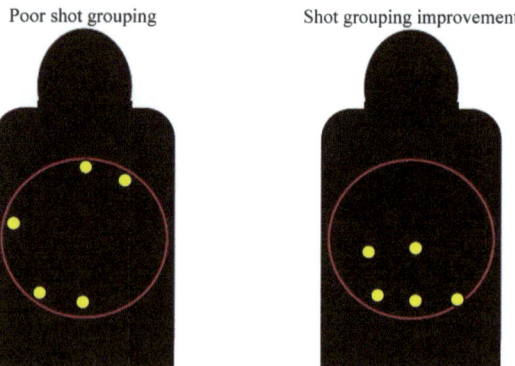

Figure 8.3. Shot group improvement

8-5.   Target analysis.

   a.   This is NOT breathing!

      1.   Remember, all firing takes place at the weapon.

      2.   The only thing you can tell for sure is that the rifle was pointed
           in those spots when it went off.

Figure 8.4. NOT breathing

   b.   Shots grouped, inconsistent elevation placement.

      1.   Perception of center mass hold not consistent.

      2.   Concentrate on maintaining consistent stock weld.

      3.   Having shooter aim at a smaller area of the target can help them
           focus and gain more consistency.

      4.   Some fixes for shooters that have trouble maintaining elevation
           holds are to highlight center elevation.

         a) Have the Shooter fire at a smaller target. A 1 ½ inch square
             will give them less target to float around in helping to isolate
             center mass and shrink group size.

         b) Another fix is to have the shooter fire at a 2 in "donut" target.
             This enables them to focus on sight alignment while still
             having a reference for sight placement.

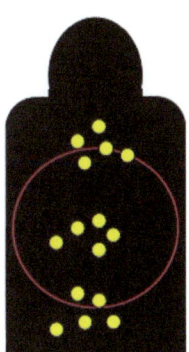

Figure 8.5. Grouped but
inconsistent elevation

c. Shots grouped, inconsistent placement.

   1. Point of aim is most likely not the problem.

   2. Proper application of principles within each shot group.

   3. Indication of a gross change in sight alignment between shot groups.

   4. Indication of poor natural point of aim.

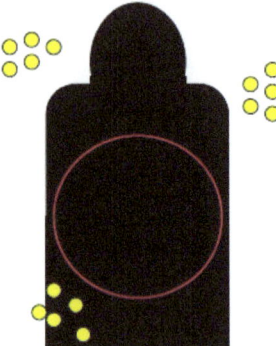

Figure 8.6. Grouped but inconsistent elevation

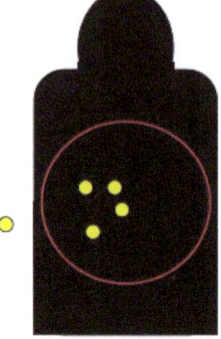

d. Bad shot in an otherwise good group.

   1. Shooter could have pointed a bad shot.

   2. Possibly poor trigger control.

Figure 8.7. Bad shot

e. No grouping.

   1. Shooter could be making one, or multiple, fundamental errors.

   2. Watch the shooter shoot.

   3. Systematically work through each fundamental task and scrutinize each shot group.

   4. Having shooters aim at a smaller area of the target can help them focus and gain more consistency.

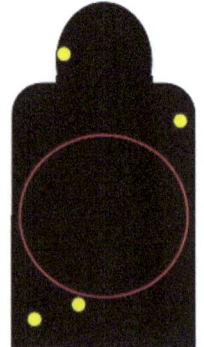

Figure 8.8. No group

42

CHAPTER 9 – COMPLEX ENGAGEMENTS

RANGE ESTIMATION

9-1.   Having an accurate range to the target is important for several reasons.

    a.   Knowing the distance to partially exposed short-range targets will allow you appropriately compensate for the weapon's Battle Sight Zero and for longer range targets, to compensate for bullet drop and wind.

    b.   Knowing ranges to targets assists in the collection of battlefield data. This can result in friendly forces ability to ascertain enemy capabilities.

9-2.   Judging distance. There are 3 main factors that affect your ability to judge distance with your eyes.

    a.   Nature of the terrain.

        1.   Objects will appear closer when looking across a depression when most of which is hidden by view.

        2.   Objects will also appear closer when looking up towards high ground; along a straight open road or railroad tracks.

        3.   Objects will appear farther than they really are when looking across a depression when all of it can be seen.

        4.   Objects will appear farther when looking from high ground to low ground when the field of view is narrowly confined.

    b.   Nature of the light. The more clearly a target can be seen, the closer the target will appear.

        1.   A target viewed in full sunlight appears to be closer than the same target viewed at dusk or dawn or through smoke, fog, or rain.

        2.   When the sun is behind the shooter the target will appear closer.

        3.   When the sun is behind the target the more difficult the target will be to see and will appear farther away.

    c.   Nature of the target.

        1.   The larger the target the closer it will appear.

        2.   The smaller the target the farther it will seem.

3. A target that contrasts with its background will appear closer.

4. A target that has an irregular outline will appear farther.

5. The easier a target is to see, the closer it will appear, the harder the target is to see or the more obscured the target is, then the target will appear farther away.

9-3. Ranging the target: Requires routine practice on known and unknown distance ranges or training areas. The Soldier should use the method that he is both most confident in and is the easiest to understand for that individual.

a. Football field method- How many football fields between you and the target.

1. Estimate 100m (football field), then determine how many of these units will fit between you and the target.

2. This method's accuracy is limited by the ground visibility.

3. Accurate to about 500m.

300 METERS

Figure 9.1. Football field method

b. Appearance of objects method – How objects look over varying terrain.

1. This method requires the viewer to be familiar with the sizes and details of personnel and equipment at certain distances.

2. Limited by visibility and familiarity of target.

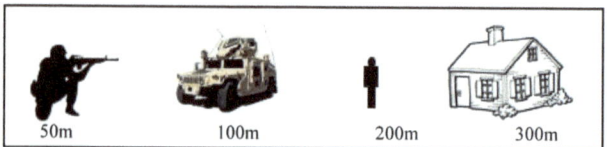

Figure 9.2. Appearance of objects method

c. Visible detail method.

    1. Observing the amount of detail on the target at various ranges indicates the distance.

    2. Eyesight dependent.

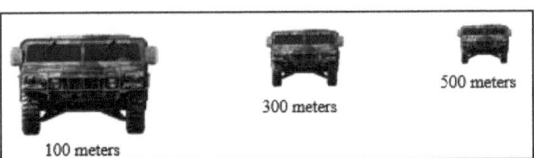

Figure 9.3. Visible detail method on Humvee

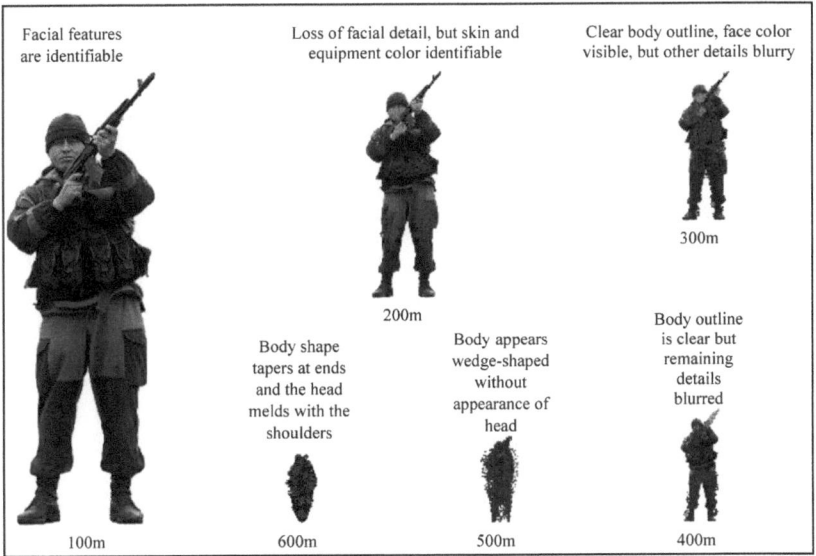

Figure 9.4. Visible detail method on human

d. Bracketing

    1. Use any method or methods of range estimation you are comfortable with.

    2. Estimate the shortest possible distance, then the farthest possible distance.

    3. The average of those distances is the estimated range to the target.

e. Map method

 1. Plot your location on map.

 2. Plot the location of the
 target.

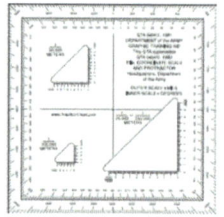

 3. Measure distance between
 the two points.

Figure 9.5. Map method

 4. Good for filling out range
 card in defensive fighting position.

f. Range Estimation with an
 ACOG

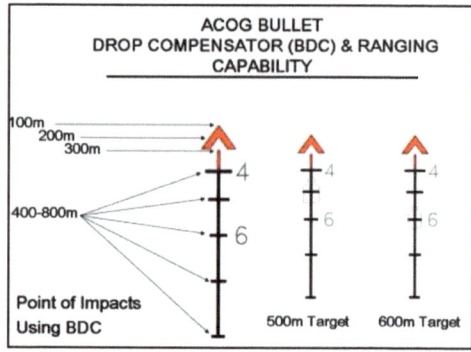

Figure 9.6. Range estimation with ACOG

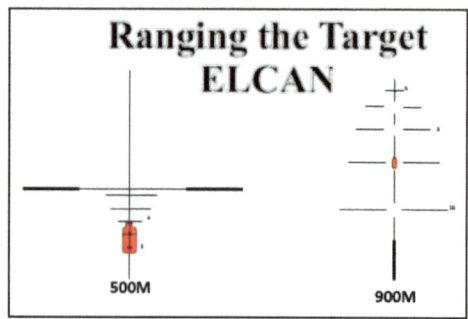

Figure 9.7. Range estimation with ELCAN

g. Range Estimation with an ELCAN

 1. 100 – 600 meter are equal to 19 in
 at the indicated range (average
 width of a man sized target).

 2. 700 – 1000 meter stadia "gap" is
 39" tall by 38" wide at the
 indicated range.

 3. 700 –1000 meter stadia lines
 indicate hold-off for a 10 mph
 wind at the indicated range

WIND

9-4.   Wind has the greatest environmental effect on the bullet. It is important to be able to read the wind in order to properly zero the weapon and engage targets at extended ranges. Wind has the greatest VARIABLE effect on ballistic trajectories.

9-5.   Elements of wind effects.

a. The time the projectile is exposed to the wind (DISTANCE to target).

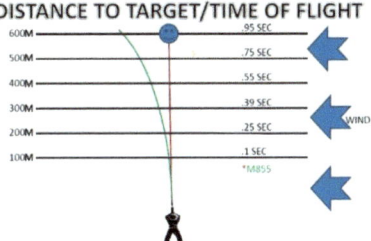

DISTANCE TO TARGET/TIME OF FLIGHT

600M	.95 SEC
500M	.75 SEC
400M	.55 SEC
300M	.39 SEC
200M	.25 SEC
100M	.1 SEC
	*M855

Figure 9.8. Distance vs time of flight

b. Direction the wind is blowing.

1. Full value wind- A wind coming directly from your left (9 o'clock) or from your right (3 o'clock). A full value wind has the most effect on a bullets trajectory.

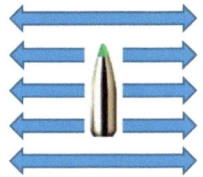

Figure 9.9. Full value wind

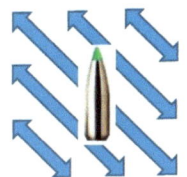

2. Half value wind- A wind coming from 1:30, 4:30, 7:30, or 10:30 o'clock is worth half as much as a full value wind.

Figure 9.10. Half value wind

3. No value wind- A wind coming from 12 o'clock or from 6 o'clock has no impact on a bullet's trajectory.

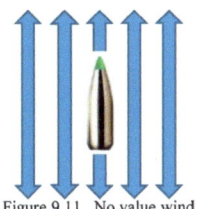

Figure 9.11. No value wind

c. Velocity of the wind.

47

9-6.   Wind Estimation.

        a.   The most important wind indicators will be the wind indicators from 1/2 to 2/3 of the way between the shooter and the target.

        b.   The wind blowing at your location may not necessarily be the same wind blowing on the way to your target.

        c.   The wind at the 1/2 to 2/3 mark to your target will have the most significant effect on your bullet since that is the point where most rounds have lost a large portion of their velocity and are beginning to destabilize.

9-7.   Wind indicators. Use visual cues from the environment to determine wind velocity and direction. Mirage, the reflection of light seen through layers of air at varying temperature, much like the heat waves you see coming off of a car roof on a hot day, can be observed using optics (binoculars, scopes, etc.). Mirage can be used to determine wind direction, and speed through observation. This skill takes time and practice to master.

    a.   Felt on skin.

    b.   Foliage movement.

    c.   Debris and dust movement.

    d.   Flags and smoke.

    e.   Mirage.

Figure 9.12. Wind indicators

9-8.   Wind speed estimation quantified. Below are some indicators to look for when estimating wind speed. Different combat environments have different wind indicators available, so use your imagination.

        1.   0-3 mph - Hardly felt, but smoke drifts.

        2.   3-5 mph - Felt lightly on the face.

        3.   5-8 mph - Keeps tree leaves in constant movement.

        4.   8-12 mph - Raises dust and loose paper.

        5.   12-15 mph - Causes small trees to sway.

9-9.   Learning to Read Wind. Wind reading takes a long time to master. To help you understand wind, observe your surroundings and check your estimation against a wind meter (anemometer). A commonly used wind meter throughout the Army is the Kestrel brand. You can shoot in windy conditions and note the location of your shots and the conditions you observed while firing them. You can begin to learn the effects of wind by observing the reaction of the bullet through optics. If you stand behind a shooter when they fire a round while looking in the direction of the target through binoculars or a scope you should be able to see the disturbance of the air cause by the bullet. Watching this disturbance you will be able to tell the path the bullet took downrange which will show you the amount of effect the wind had on the bullet.

9-10.   Adjusting for wind. Once you've estimated the wind velocity and direction, now it's time to estimate your hold off or rear sight windage correction. This is accomplished by using the simplified wind formula. This formula is only for use with M16/M4 type weapons using 5.56mm ammunition. This formula will allow you to adjust for the distance that the wind displaces your bullet.

a.   Take your range estimation in meters and divide by 100. For example 200m/100 would equal 2.

b.   Multiply that result by the wind velocity in MPH.

c.   Divide that result by 7 and you get your wind drift in minutes of angle (MOA).

d.   The result is for a full value wind. If the wind is half value, divide the result by 2.

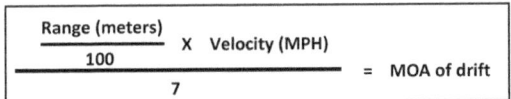
Figure 9.13. Simplified wind formula for 5.56mm

e.   Example: the estimated distance is 400m and wind velocity is 7MPH from 4 o'clock.

1.   400m/100 = 4.

2.   4 x 7 = 28

3.   28 / 7 = 4

4.   4 / 2 = 2 (half value wind)

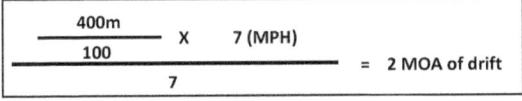
Figure 9.14. Simplified wind formula example

9-11.   Bracketing your wind condition. Developing a bracket for the conditions you are shooting in is proactive rifle marksmanship. It takes little to no time to do and will increase the likelihood of hitting your target sooner.

   a.   Estimate maximum and minimum wind velocities.

   b.   Estimate value of wind (full, half, no).

   c.   Calculate wind holds for maximum and minimum.

   d.   As conditions vary, adjust hold by referencing maximum and minimum.

9-12.   Zeroing in Wind. If the sights are dialed in center of the target while there is a wind condition, the zero will be off. This is because the weapon was zeroed for a particular wind condition and will only hit center during that same wind condition. It is important to know what the wind is doing so that we may zero properly.

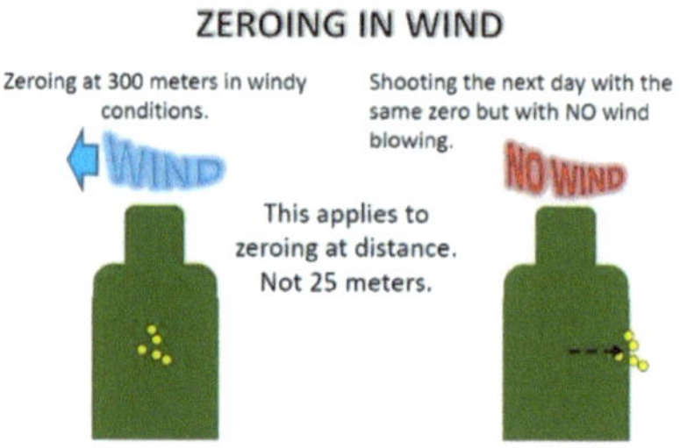

ZEROING IN WIND

Zeroing at 300 meters in windy conditions.

Shooting the next day with the same zero but with NO wind blowing.

WIND

NO WIND

This applies to zeroing at distance. Not 25 meters.

Figure 9.15. Zeroing in the wind at distance

9-13.   Other Weather Factors. Although wind has the greatest effect on your bullet trajectory there are other factors to take in to account.

   a.   Temperature – temperature affects muzzle velocity. Hotter temperature equals faster powder burn rate which equals faster muzzle velocity. This may affect your zero.

   b.   Temperature – temperature affects air density. Cool air is thicker and will slow your bullet down faster. If you zero in a dense atmospheric condition and move to a thin air condition, your zero may change.

   c.   Altitude – altitude affects air density as well. If your weapon was zeroed at a low altitude and you are operating in at a high elevation there may be a change to your zero.

Bottom Line - If you zero in one location or condition and then go to a drastically different one…reconfirm your zero.

MOVING TARGETS

9-14.   Targets on the battlefield are often in motion. In order to successfully engage a moving target the shooter must know how far ahead of the target to aim. This is called using a LEAD, or LEADING the target. Factors involved in determining moving target leads:

    a.   Time of flight of the round.

    b.   Speed of target.

    c.   Direction of movement.

    d.   Wind.

9-15.   Leads. It is important to reference your leads from center mass of the target, rather than the leading edge. This assures the impact of the round is not effected by the size and/or angle of the target. This will also result in more consistent leads and creates better sight picture for maximizing accuracy. Figures 9.16 and 9.17 are for the speed of a walking man.

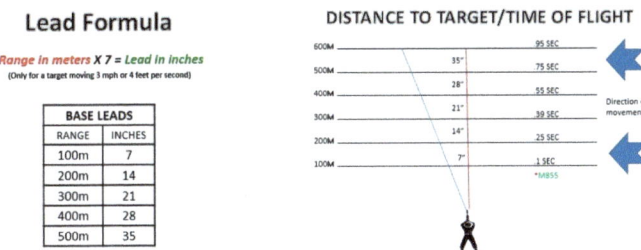

Lead Formula

Range in meters X 7 = Lead in inches
(Only for a target moving 3 mph or 4 feet per second)

BASE LEADS	
RANGE	INCHES
100m	7
200m	14
300m	21
400m	28
500m	35

Figure 9.16. Lead formula

DISTANCE TO TARGET/TIME OF FLIGHT

600M			95 SEC
500M		35"	75 SEC
400M		28"	55 SEC
300M		21"	39 SEC
200M		14"	25 SEC
100M		7"	1 SEC
			*MISS

Direction of movement

Figure 9.17. Time of flight

For other targets the complete moving target lead formula is:

    a.   Time of flight of the round x Target speed (feet per second) = Lead in feet.

    b.   Lead in inches / 3.6 inches = mil lead.

9-16.   Oblique Movers

    a.   When engaging a 45 degree moving target use half the lead.

    b.   Half the lead is needed because the target is only covering half the ground.

    c.   Targets moving directly toward or directly away do not need to be led.

9-17. Moving target engagement methods:

a. Trapping – The trapping method is used with the rifle stationary and sights settled. The target is engaged when the target moves into the predetermined point. The trigger is pulled simultaneously with the establishment of the proper lead and sight picture.

Figure 9.18. Trapping

b. Tracking - The tracking method is used for a moving target that is progressing at a steady pace over a well-determined route. The target is tracked with the weapon's sights while maintaining a point of aim on or ahead of (leading) the target until the shot is fired.

Figure 9.19. Tracking

c. Moving target engagement process.

1. Identify target

2. Establish position

3. Compensate for distance to target, wind, and speed of target

4. Aim

5. Fire/Follow through

6. Call your shot

7. Reassess target/Re-engage target

OTHER CONSIDERATIONS

9-18.   Target to target transitions.

    a.   Width. After engaging one target, eyes move to next target either to the right or left of previous target. Shooter should drive the weapon with their non-firing hand so their sights meet up with their eyes on the next target. Once the appropriate sight picture has been achieved, shooter should apply proper trigger squeeze and engage the target. Shooter should be careful not to dip the muzzle down or place the weapon on safe as these are both unnecessary and waste time. Targets can be shot at the same speed.

    b.   Depth. After engaging one target, eyes move to next target either closer or farther than the previous target. Shooter should drive the weapon with their non-firing hand so their sights meet up with their eyes on the next target. Once the appropriate sight picture has been achieved, shooter should apply proper trigger squeeze and engage the target. Shooter should be careful not to dip the muzzle down or place the weapon on safe as these are both unnecessary and waste time. The closest target can be shot more quickly than the farther target while still achieving the same lethal hits.

9-19.   Quick High Angle Formula. For shooting up or down slope at extreme angles. The three things that must be known to use the quick high angle formula are the distance to the target, the hold for that given distance and the angle to the target. These may have slight errors but are extremely close and will get you on target quickly.

    a.   30 deg – subtract .5 MOA for every 100m from the original hold.

    b.   45 deg – subtract 1 MOA for every 100m from the original hold.

    c.   60 deg– subtract 2 MOA for every 100m, then add 1 MOA back from the original hold.

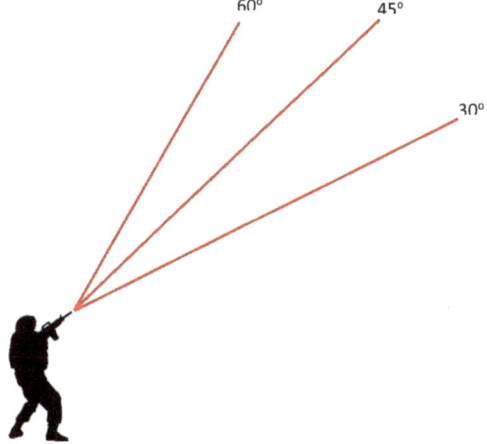

Figure 9.20. High angle

CHAPTER 10 – AMMUNITION

10-1.   M855 – Ball (green tip): Currently the most commonly issued 5.56 round. Intended for personnel and unarmored targets. **Note** during training, this round CANNOT be used to engage steel targetry due to the over penetrating capability of the round.

Figure 10.1. M855 Ball

10-2.   M855A1 – Enhanced Performance Round (EPR): Intended to replace the M855 as the general purpose round. Ability to engage steel targetry will still remain nonexistent. The EPR has a greater terminal effect on personnel targets due to its greater velocity. Primary difference between the two rounds is that the EPR has an unpainted steel tip, over a copper alloy core and a partial copper jacket. For planning training, the increased velocity of this round greatly increases the original SDZ for the green tip from approximately 30 degrees to 60.

Figure 10.2. M855A1 EPR

10-3.   MK262 – 77 gr match: Originally manufactured for the Special Purpose Rifle (SPR) and is more accurate at distance than the standard M855.

Figure 10.3. MK262 77gr Match

10-4.   M856 – Tracer (red tip): Contains a pyrotechnic composition in the base of the bullet to permit visible observation of the bullet's flight path.

Figure 10.4. M856 Tracer

10-5.   M995 – Armor piercing (black tip): An armor piercing cartridge intended for use against personnel and lightly armored targets. It is composed of a metal jacket and a hardened steel alloy core.

Figure 10.5. Armor Piercing

10-6.   AA 68 or AB 67 - Short Range Training Round (SRTR): AA 68 is composed of a blue, fluted tip, plastic projectile with an approximate range of 25m. AB 67 is composed of a copper, nylon, and carbon fiber compound likewise with a fluted blue tip.

Figure 10.6. SRTR

10-7. M200 – Blank: Intended for simulated fire during training maneuvers or ceremonial services. The "tip," of the round is crimped.

10-8. Close Combat Mission Capability Kit – (Sim Round) AB09 (blue tip) AB10 (red tip) AB11 (yellow tip): A munition intended to enhance force on force training. A special bolt carrier group is required to operate this round, however a different upper receiver/barrel is not required as the round is chambered in 5.56. **Safety Note** at the conclusion of any training event utilizing this ammunition, leaders must ensure that all barrels are clear of any obstructions.

10-9. M199 – Dummy: Primarily intended for simulated fire, and practice loading and unloading weapons during weapons handling drills.

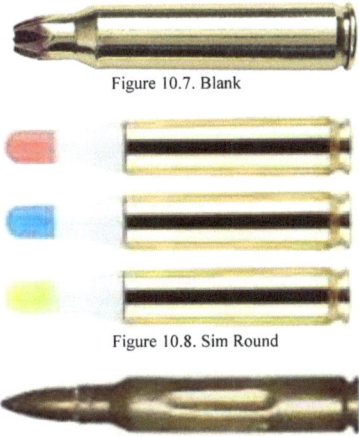

Figure 10.7. Blank

Figure 10.8. Sim Round

Figure 10.9. Dummy Round

CHAPTER 11 – MARKSMANSHIP MYTHS, MISCONCEPTIONS & FREQUENTLY ASKED QUESTIONS

The following myths and misconceptions are in no order of precedence. All of them, however, are the most commonly reoccurring issues that we have seen both while conducting our mobile training missions, and through our on-line question and answer forum. While some of these topics may seem elementary to some people, one would be surprised at the frequency in which these questions occur. The following examples are provided in a question, or statement format, with the correct answer given below:

11-1. Statement: "A 200M zero is the best battle sight zero one could have on their rifle because most combat engagements occur inside of 225 meters."

Answer: A 200M zero may not be a bad idea if you find yourself in an urban environment, but can be counter-productive when the potential for longer distance engagements is present. A 200M zero increases the probability of hits on target when shooting at closer ranges because you have lowered the maximum ordinate (trajectory) of the round, but can be a detriment if you have to engage targets beyond 200M because you will likely hit below your point of aim. This consideration becomes especially important when engaging partially obscured targets, such as an enemy combatant firing from a covered position. Additionally, when using a 200M zero, the ballistic drop compensator (BDC) is no longer calibrated, and some of your weapons functionality has been degraded. Should the need arise where you must take a further shot, you no longer have the ability to accurately make adjustments to the rear sight, and your only option is to aim over target and hope that you are successful. Ultimately, Commanders must decide what type of zero best suits their unit's needs.

11-2. Statement: "I am an Infantryman and will be issued an ACOG or M-68 for my rifle. I don't need to train with, or learn how to use iron sights."

Answer: Although you may be issued an optical sight for use with your weapon, you must still master the use of the iron sights. Should your optic become inoperable or you find yourself using other than your assigned weapon, you will still be able to successfully engage targets. Nearly all shoulder fired weapons in the US Army's inventory, as well as the weapons of most other countries, come equipped with iron sights. Additionally, knowing how to apply the fundamentals of marksmanship with iron sights will make you much more proficient when using optics. Unfortunately, this doesn't work in reverse; one who is proficient with optics will not necessarily be successful when using iron sights.

11-3. Statement: "I went to the 25M range, and my rifle is zeroed."

Answer: How do you know? Did you check it at actual distance to see where the bullets were impacting? There is a big difference between a "nearo", and a zero. If you zero your rifle to Army standard, and place 5 of 6 shots in the 4cm circle, then you are considered to be zeroed. What you may not realize is that the 4cm circle at 25M is the equivalent to a 18.5" circle at 300M; that is nearly the size of the silhouette at 300M. Let's pretend that you weren't very critical about your zero while you were at the 25M range, and although you received a "go", the majority of your shot group was off to one side of the 4cm circle. Taking into account that an angle increases over distance, when you go to the qualification range, you have less than a 50% chance of hitting a 300M target therefore, you should always confirm your zero at actual distance.

11-4. Statement: "I have my rifle zero recorded in my notebook so I can put it on any rifle that I use."

Answer: Zero's don't work like that, and rifles are mechanical in nature. When you zero a rifle, you are aligning the sights with the barrel, and the barrel with your eye (by looking through the sights). The zero that works for one rifle isn't going to work on another. Think of it like this: the same seat and mirror settings you use when driving your Ford pick-up truck will not work when you are driving your friends Volkswagen Beetle.

11-5. Statement: "The sights on all of the M-16 / M-4 family of rifles are the same, and they are all zeroed in the same manner."

Answer: They all have a different thread pitch on the elevation and windage screws, and these differences require slightly different sight settings to be used during the zeroing process. When zeroing, ensure that you are utilizing the zero targets made for the type of rifle you are using, and always reference the instructions that are printed on the bottom of the 25M zero target. These instructions are very specific on where the BDC should be set, and they differ depending on the rifle type.

11-6. Question: "Is there a noticeable shift in impact or zero from M855 and M855A1 ammunition?"

Answer: Many Soldiers have reported that they experienced a drastic shift in impact when switching from M855 to M855A1, while others reported having no significant change to their zero. The safe bet and best practice is to always zero with the type of ammunition that you will be using during training or combat. Should you need to change the type of ammunition you are using for some reason, you should reconfirm your zero as soon as possible, and then follow up with confirmation at actual distance.

11-7. Question: "When I move the sights (either front or rear) on my rifle, what happens internally to make the strike of the round move downrange?"

Answer: Nothing happens internally when you make sight corrections. The rifle is a machine, and small adjustments to the sights actually make the operator adjust their point of aim in very small increments. EXAMPLE: When you turn the BDC on your rifle clockwise, the rear sight moves up. When you look through the rear sight aperture, and reacquire sight alignment, you have simply elevated the muzzle of the weapon, thus causing the strike of the round to be higher.

11-8. Statement: "Wind has no effect on the strike of the round because bullets move entirely too fast for the wind to have any effect."

Answer: Wind, temperature, and humidity all have an effect on the strike of a bullet, with wind having the biggest effect of all. As a bullet leaves the barrel of the rifle, gravity begins to have an instant effect. As the bullet begins to run out of energy and slow down, wind has an even greater effect. As a general rule, the slower the velocity of the wind, or the shorter the distance you are shooting, the effects of wind become less significant. The higher the wind speed, or the farther out you shoot, you have to compensate for the effects of wind or you will surely miss your desired target.

11-9. Question: How is combat marksmanship different from plain marksmanship?

Answer: Yes, you may shoot at steel targets, add scenario type exercises, and increase stress, but the fundamentals are not different. "Running and Gunning," type training is a necessary and important training tool, but before all of that takes place, one must master fundamental marksmanship. Failing to master the fundamentals before graduating to scenario based exercises will only result in you shooting misses at a faster rate.

11-10. Question: "I want my unit to zero their rifles at 200M. Will cutting the 200M target out of the ALT C qualification target and using it at 25M work?"

Answer: No, it will not. While zeroing your rifle, you aim center mass and adjust the strike of the round to hit where you are aiming (point of aim, point of impact). The size of the target has nothing to do with the process, but is only there to give you something to aim at; it is your reference point. To achieve a 200M zero, you must change the distance in which you are zeroing at, or adjust your rifle to hit somewhere other than where you are aiming.

11-11. Statement: "In order to achieve a 200M zero, you must zero at 37M."

Answer: Given the weapon systems, ammunition, and zero targets available in the Army's inventory, this is impossible. In order to achieve a true 200M zero with an M16A2, A3, or A4 you would have to zero the rifle at 50M. To achieve a true 200M with an M4, it would need to be zeroed at 41M. The best practice for achieving any zero, regardless of the desired distance, is to obtain an initial zero at 25M, then fine tune your zero at the actual desired distance.

11-12. Statement: "When I am teaching marksmanship, I always make my Soldiers train in IBA because that is how they are going to fight."

Answer: Learning the fundamentals requires an undistracted mind and an unstressed body. IOTV, IBA, kit interferes with those conditions. Only once a Soldier can routinely execute the fundamentals should the level of difficulty increase by integrating individual combat equipment into dry firing and live fire exercises. This also assists NCO's in identifying what may be going on with Soldiers who are having trouble.

11-13. Statement: "He is shooting all over the place. I told him to watch his breathing."

Answer: Entirely too much emphasis has been placed on breathing, and breathing probably has very little to do with the most problems that are observed on the range. A person firing a rifle has a tendency to fire the rifle while they are in their natural respiratory pause, and the rifle isn't moving. Although discussing breath control and the natural respiratory pause should be included in any marksmanship instruction, don't over emphasize it. It is impossible to determine what a shooter may be doing wrong by analyzing a shot group on a piece of paper. The only way to accurately determine what is going wrong is through **CAREFUL** observation of a peer coach who is actively watching the shooter throughout the entire firing process.

11-14. Statement: "I always make my Soldiers put the tip of their finger on the trigger, and the tip of their nose on the charging handle. That is the way I was taught."

Answer: Due to the fact that we are all made differently, the "cookie cutter" approach to shooting positions is not the best approach when teaching others. There are some key points to consider when teaching shooting positions: 1) A person has to be reasonably comfortable to make well aimed shots. 2) A Soldier's shooting position, regardless of which position it is, must allow them to achieve proper sight alignment consistently from shot to shot. 3) The position of the firing hand is critical in controlling the rifle, applying proper trigger control, and managing recoil for follow up shots. The firer's hand position should be high on the pistol grip, and their finger should naturally lie on the trigger. Finally, the trigger is a lever. The trigger must be moved straight to the rear. Forcing someone to place their finger on the trigger in an unnatural position will lead to poor trigger control habits, and misplaced shots downrange.

11-15. Question: "After we zero our rifles and complete training, our unit armorer takes our carrying handles/optics off of our weapons because they don't fit in the gun racks with them on. Does this affect my zero?"

Answer: Maybe. Manufacturers of the 1913 rail system that most military rifles come equipped with do not guarantee that you will not experience a shift in impact after removing and reinstalling your optic. After removing any part of the weapon's sighting system, whether using an optic or iron sights, a weapon should be zeroed again before any live firing. Although you may get lucky sometimes, and find that your point of impact did not move all that much, it is not uncommon for rifles to drastically off target after having the sights removed.

This Supplemental is intended to expound upon the information found in TC 3-22.9. The information has been provided by United States Army Marksmanship Unit and has been approved for release by United States Army Infantry School within the Maneuver Center of Excellence. USAMU is part of the U.S. Army Accessions Brigade and Army Marketing and Research Group.